PEEING and POOPING in
SPACE

PEEING and POOPING in SPACE

A 100% FACTUAL ILLUSTRATED HISTORY

KIONA N. SMITH

ILLUSTRATED BY HEADCASE DESIGN

RUNNING PRESS
PHILADELPHIA

Running Press
Hachette Book Group
1290 Avenue of the Americas, New York, NY 10104
www.runningpress.com
@Running_Press

First Edition: May 2024

Published by Running Press, an imprint of Hachette Book Group, Inc. The Running Press name and logo are trademarks of Hachette Book Group, Inc.

The Hachette Speakers Bureau provides a wide range of authors for speaking events. To find out more, go to www.hachettespeakersbureau.com or email HachetteSpeakers@hbgusa.com.

Running Press books may be purchased in bulk for business, educational, or promotional use. For more information, please contact your local bookseller or the Hachette Book Group Special Markets Department at Special.Markets@hbgusa.com.

The publisher is not responsible for websites (or their content) that are not owned by the publisher.

Print book cover and interior design by Katie Benezra.

Library of Congress Cataloging-in-Publication Data has been applied for.

ISBNs: 978-0-7624-8661-8 (hardcover), 978-0-7624-8662-5 (ebook)

Printed in China

APS

10 9 8 7 6 5 4 3 2 1

TO MOM, ALWAYS AND FOREVER;
FAMILY ABOVE ALL.

TO DADDY, WHO TOLD ME THIS
WOULD HAPPEN.

CONTENTS

INTRODUCTION: THE LAST TOILET ON EARTH • IX

CHAPTER ONE: GOING NUMBER ONE 1

CHAPTER TWO: EVERYBODY POOPS 17

CHAPTER THREE: SPACE TOILETS 29

CHAPTER FOUR: UNDER WHERE? 47

CHAPTER FIVE: **IT HAS TO GO SOMEWHERE** 59

CHAPTER SIX: **FARTING AROUND** 71

CHAPTER SEVEN: **OUT THE WRONG END** 83

EPILOGUE: **HANGING UP THE FUNNEL** 93

A TIMELINE OF PEEING AND POOPING IN SPACE • 96

SUGGESTED FURTHER READING • 97

ACKNOWLEDGMENTS • 98

THE LAST TOILET ON EARTH

High on the tower at Launch Pad 39A at NASA's Kennedy Space Center, 195 feet above the Florida coast, is a restroom with a view. The extremely spartan pit stop features a cold metal toilet, a tiny sink, and some obscene graffiti allegedly written by astronaut John Glenn in 1998. Here, astronauts stopped for a last normal trip to the bathroom before boarding the Space Shuttle; today, the barebones restroom is a last outpost for SpaceX Crew Dragon crewmembers. It's nicknamed the "Last Toilet on Earth."

This book is about what astronauts do when they're hundreds of miles beyond the Last Toilet on Earth and still need to go.

How astronauts go to the bathroom in space is "the all-time favorite question [we] are asked," according to Apollo 11 pilot Michael Collins. During the years of the two-person Gemini missions (flown from 1965 to 1966) and the three-person Apollo missions (flown from 1968 to 1972), almost no one was better equipped than Collins to answer that question. Besides his hands-on experience, he also played a role in designing the waste disposal systems in the astronauts' pressure suits.

The answer, Collins gleefully told a reporter in 2019, is "Carefully."

If you're the kind of person who wants to know more, you've picked up the right book. The more detailed answer to Collins's all-time favorite question involves equal parts engineering, comedy, and daredevil adventure. It's a story about technology, but it's also a story about figuring out how to be human in a completely new environment.

In fact, using the bathroom in space is a decent metaphor for life in general: it's awkward, it takes courage and practice, and it gets a little easier if you can learn something and laugh about it.

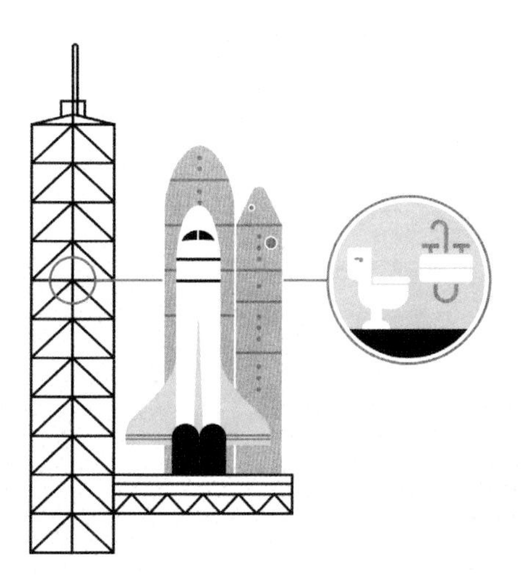

CHAPTER ONE

GOING NUMBER ONE

Nobody expected the first astronauts to need to go.

The first flights to space were very short. Yuri Gagarin's first orbit around Earth lasted just over an hour from launch to splashdown, and Alan Shepard's suborbital flight took around fifteen minutes. Even with a few hours on the launchpad for checklists and countdowns, both American and Soviet mission planners figured their brave space explorers could simply hold it.

But one of the first rules of spaceflight is that things don't always go according to plan.

YURI GAGARIN'S PIT STOP

On April 12, 1961, Soviet cosmonaut Yuri Gagarin—the first person to fly to space—felt the call of nature before he was suited up and strapped into his Vostok 1 capsule. Gagarin did the most practical thing he could think of: he relieved himself on the right rear tire of the bus that had come to drive him to the launchpad located at the Baikonur Cosmodrome.

Spacecraft still launch cosmonauts and astronauts from the Baikonur Cosmodrome, and buses still carry them from their quarters to the launchpad. And the bus still stops halfway so crews can pile out and ceremonially pee on the right rear tire, in honor of Gagarin's historic pit stop. Male astronauts use their own equipment, while women splash a cup of urine or water onto the tire. Even foreigners, like American astronaut Scott Kelly, have participated.

The splashy preflight tradition may not last much longer, however. In 2019, the Russian space agency, Roscosmos, showed off a prototype for a new model of spacesuit, the Sokol-M. When—or if—the Sokol-M replaces today's Sokol-K suits, cosmonauts won't be able to perform the roadside reenactment. Instead of a typical fly zipper, the new suit's zipper cuts diagonally across the torso, which looks cool but makes it impossible to unzip for a quick go.

"We have the design specifications. They don't state that it's necessary to pee on the wheel," the director of Zvezda, the company that designed the prototype, told the press back in 2019. "The design specifications would need to be adapted."

"JUST GO IN THE SUIT"

Astronaut Alan Shepard—the first American in space—didn't think to make a Gagarin-style pit stop on the way to the launchpad at Cape Canaveral, Florida, on May 5, 1961. After waiting in his Mercury capsule for several hours more than he'd expected, eventually Shepard realized he had a problem. It's easy to imagine him staring longingly at the "liquid waste" container near the entrance hatch of his Mercury capsule. Strapped into his flight seat, Shepard couldn't possibly reach the container, let alone use it without shedding his bulky pressure suit. (Engineers had installed the container mostly for the sake of appearances. After all, nobody expected an astronaut on a fifteen-minute suborbital flight to need it.)

After some back-and-forth with mission control, Shepard received a historic set of instructions: "Just go in the suit."

And he did. A short while later, Shepard rode his Redstone rocket to space in a soggy pressure suit, with pee short-circuiting the heart and respiration sensors attached to his body. The suit is now on display in the National Air and Space Museum. Hopefully it's been cleaned.

THE FIRST TO GO NUMBER ONE IN ORBIT

In February 1962, a few months after Shepard's slightly soggy flight, astronaut John Glenn became the first American to orbit Earth. Before returning to Earth, Glenn also became the first person to actually pee in space (that we know of; the Soviet space program kept a lot of details quiet).

Luckily for Glenn, NASA had learned from Shepard's discomfort—and from the design of pressure suits for pilots of high-altitude reconnaissance planes like the U-2—and gave every astronaut after Shepard a way to relieve themselves. Beneath the pressure suit, Glenn wore a latex bag fitted over his, ahem, personal equipment. He could pee whenever the need arose without fearing for his medical sensors (or his spacesuit).

Glenn's need arose fairly late in the flight, and that caused him to set yet another, even less glamorous, record. The average human bladder holds about twenty ounces of fluid, but Glenn gave the space program his all—which in this case meant about twenty-seven ounces.

It turns out that floating in space makes it harder to tell when you need to pee. Here on Earth, gravity causes fluid to settle in the bottom of the bladder, where its weight eventually cues nerves lining the bladder that it's time to go. In orbit, fluid just floats around the bladder, and the need to go only kicks in when the sides of the bladder start stretching. By then, the bladder is *very* full, as Glenn discovered. Some astronauts have reportedly peed on a schedule to avoid risking permanent damage to their bladders.

ONE SMALL INDIGNITY FOR A MAN . . .

The Mercury missions—including Shepard and Glenn's flights—proved we could send people into space. Starting in 1961 the Gemini program aimed to prove people could stay up there longer and do more while they were at it. Most of the Gemini crews spent one to four days in orbit. Astronauts Frank Borman and Jim Lovell spent a record-setting two weeks in space during their Gemini 7 mission in 1965. And astronauts living in space for days at a time were *definitely* going to need to relieve themselves.

NASA's solution was the Urine Collection Device: a roll-on latex cuff, like a condom with the end cut off. A tube with a valve and clamp connected the cuff to a latex bag which, like John Glenn before them, the Gemini astronauts wore strapped to a belt under their pressure suits.

The belt and bag stayed on for the entire multi-day mission; the cramped Gemini capsules left no room to extract one-self from a bulky pressure suit. Instead, astronauts like Michael Collins (who would later fly the Apollo 11 crew to the Moon) could unzip their suits, roll on the cuff, connect

it to the bag, and fire away from the comfort of their seat—the same seat in which they ate and slept.

During launch and reentry, or when venturing outside the capsule for an extravehicular activity (better known as a spacewalk), astronauts obviously couldn't just unzip their pressure suits, so for those activities, they had to roll the cuff on and leave it there, underneath the suit.

SIZE DOES MATTER

When it came to the roll-on cuffs, size was critical. If the cuff didn't fit properly, it could leak or even come off mid-stream, sending little globules of pee floating around the cabin. And during ground tests, when Gemini crews sat for hours in a cramped capsule with their feet above their heads, the wrong cuff size could leave urine running down their backs.

The system had been designed to fit male anatomy, since all astronauts in the 1960s (and until 1978) were men, but NASA didn't plan for

EXTRA LARGE

IMMENSE

UNBELIEVABLE

the fact that all its Gemini astronauts were also test pilots—brave, swaggering, larger-than-life heroes with the egos to match. Naturally, they all wanted to claim the large cuff size. They were not always correct.

NASA developed its own version of vanity sizing, with the full cooperation of the astronauts: they renamed the "small, medium, and large" cuffs in what Collins later described as "more heroic terms: extra large, immense, and unbelievable."

Reportedly, engineers briefly considered molding each astronaut's anatomy to produce a custom-fit cuff, but the astronauts themselves balked at the idea. The National Air and Space Museum's Space History Department curator, Jennifer Levasseur, speculates that most of them didn't want their actual sizes on record.

And she may be right; Space Shuttle astronaut Michael Mullane wasn't fond of NASA records reporting which size he selected of the four available in 1980.

"I don't care if they publish my medical records in the *New York Times*," he told a colleague. "I just hope the record of my condom size is locked up in a vault in Cheyenne Mountain."

IN-FLIGHT BATHROOM READING

Besides dealing with the roll-on cuff, astronauts on the Gemini missions had to take urine samples every time they peed so NASA could study the levels of electrolytes and other important chemicals involved in keeping the body hydrated. NASA sent along a twenty-step set of instructions to follow, of which actually peeing was the fourth item. Items five through twenty involved taking a sample—for science!—emptying the rest of the urine overboard, stowing the sample, and properly cleaning and putting away the cuff. (NASA checklists are nothing if not thorough.)

That left Gemini crews with plenty of bathroom reading material. But what they didn't have was privacy. Candidly, Collins noted in his memoir that he was relieved not to have to suffer the indignities of the Urine Collection Device in mixed company. It was bad enough, he wrote, doing it in front of his Gemini 10 crewmate, John Young.

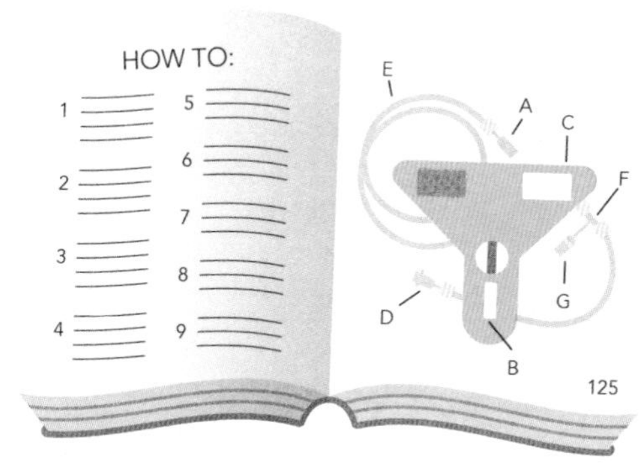

. . . ONE GIANT LEAK FOR MANKIND

In the late 1960s, the Apollo program followed right on the heels of Gemini, and its goal was clear but far from simple: land astronauts on the Moon. And within moments of stepping off the ladder of the Lunar Module onto the regolith (i.e., moon dust) of Tranquility Base on the Moon, with the whole world watching via the television camera mounted to the outside of the Lunar Module, astronaut Buzz Aldrin realized he needed to pee.

“Neil [Armstrong] might have been the first man to step on the Moon, but I was the first to pee his pants on the Moon,” Aldrin later wrote in his autobiography. “I was, of course, linked up with the urine-collection device, but it was a unique feeling. The whole world was watching, but I was the only one who knew what they were really witnessing.”

CHAPTER TWO

EVERYBODY POOPS

Alan Shepard's 1961 launchpad plight could have been much worse; at least he only had to go "number one."

As the popular children's book reminds us, everybody poops, even astronauts in space. But microgravity makes everything more complicated. Without gravity to pull things downward, fluid tends to stick to skin and clothes thanks to a force called surface tension, which creates problems you probably don't want to think about when it comes to using the bathroom. And the human body also takes some time to adjust.

"The bowel does not at first function normally with the intestines floating inside the weightless astronaut," explains Shuttle astronaut Jerry Linenger in his autobiography. Even using the modern toilets aboard the International Space Station (ISS), astronauts sometimes struggle to get the job done on their first few tries.

In the 1960s and early 1970s, before NASA launched the first space toilet aboard its Skylab space station (see Chapter Three), pooping in space definitely wasn't for the faint of heart, bowel, or stomach.

ASTRONAUTS' PREFLIGHT ANTI-POOP DIETS

The Mercury missions—the first six crewed flights into space—were all fairly short. In May 1963, astronaut Gordon Cooper spent almost a day and a half on his Mercury-Atlas 9 flight, which was longer than the first five Mercury flights combined. Most of the subsequent Gemini flights lasted less than four days (with two exceptions). On such short flights, most astronauts could avoid the "number two" issue altogether with some careful planning (the Gemini 7 crew were stuck with it, however).

Nearly everyone ate a low-residue diet in the days before a launch. That meant avoiding whole grains, fruits with seeds or peels, high-fiber vegetables like broccoli and greens, and even crunchy peanut butter. Even the traditional prelaunch breakfast of steak and eggs was a high-protein, low-residue meal designed to make astronauts have fewer—and smaller—bowel movements.

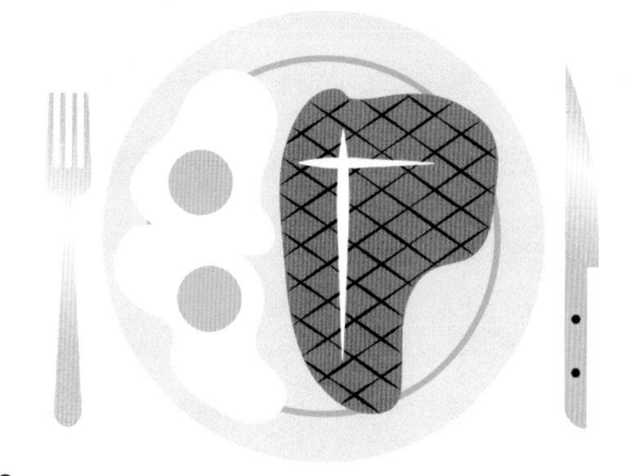

The astronauts' goal was to get through their two- or three-day flights (like nearly all of the Mercury missions and all but two of the Gemini missions) without having to poop. On longer missions, like later Apollo flights, they hoped to at least put it off as long as possible.

Apollo 16 astronaut Charlie Duke told *Vox*'s Brian Resnick in 2019, "Three days is not bad to have without a bowel movement," and that's technically true (but please don't try it at home).

Even in the age of the modern space toilet, some astronauts take similar precautions; Linenger reported that he had three glasses of (pulp-free) juice for breakfast the morning of his launch to the Russian space station Mir in 1997, and some of his fellow astronauts would eat a soft, low-residue diet for *weeks* before launch. Their goal was to give their bodies a few days to adjust to microgravity before they had to do their business.

PRELAUNCH LAXATIVES AND OTHER DRASTIC MEASURES

Some of the Apollo astronauts went to great lengths to avoid using the in-flight facilities. They opted to take some prelaunch laxatives and get everything out of their systems before the flight. Once in flight, some even resorted to anti-diarrhea medicines to slow things down.

NASA engineers quickly learned all about the careful low-residue diets, the laxatives, and the anti-diarrhea pills, and those tricks found their way into an official report on the successes and failures of the Apollo program's waste disposal efforts. Its authors had to admit that astronauts were, frankly, grossed out about pooping into NASA's carefully engineered receptacles.

Soviet cosmonauts took it a step further and developed a tradition even more bizarre than Gagarin's famous pit stop: a pre-flight enema. It was still in full swing by the time Scott Kelly caught a ride to the ISS aboard Russia's Soyuz capsule in 2015.

"The cosmonauts have their doctors do this," Kelly wrote. "But I opt for the drugstore type in private."

HOW TO POOP ON THE WAY TO THE MOON

Giving their guts a rest wasn't the only reason Gemini and Apollo crews tried to stave off bowel movements in space. The facilities were awkward . . . and awkwardly hands-on.

NASA devised "an extremely basic system" for keeping astronauts' personal business contained. Feces had to go directly from the astronaut into a sealable container, with no chance of escaping and wandering around the command module. The solution was the

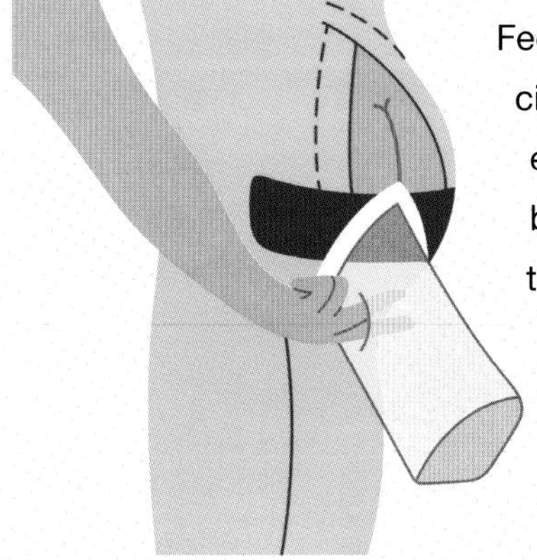

Fecal Collection Assembly, a very official-sounding name for something even NASA described as "a plastic bag taped to the buttocks to capture feces."

To use the bag, astronauts opened the back flap on their "Constant Wear Garments," which looked a bit like long underwear (only with short sleeves). Not having to actually pull their pants down was the only convenient part of the process, however. An astronaut needed to literally tape the flanges on the plastic bag to his bottom to prevent anything from floating away. But the Constant Wear Garment's access flap was just slightly too small for the fecal containment bag, so the process was tricky at best.

It got worse from there. We take for granted that when we do our business, it will fall away from our bodies and into the toilet. In space, it needs a little help. The bag included a little pouch near the top, just the right size for a finger; astronauts had to reach in, using the finger-pouch, to detach each piece of feces and nudge it toward the bottom of the bag. The preflight purges suddenly didn't seem too bad by comparison.

Once the job was done, the astronaut had to quickly clean himself with tissue wipes, open a pouch of antibacterial liquid, toss it and the wipes into the bag, and seal the whole thing. The astronaut—who had spent months training to fly the most complicated machine humans had ever built—would spend the next several minutes kneading the bag to mix the feces and the antibacterial liquid together. Once that indignity was complete, it was time to roll the bag up as small as possible and put it in the waste storage compartment.

The whole process took about forty-five minutes, which astronaut Walter Cunningham painstakingly recorded in Apollo 7's logbook.

MOON DIAPER, NEVER USED (ALLEGEDLY)

Astronauts couldn't open a handy back flap on their spacesuits while walking around on the Moon, so they needed another option: the Fecal Containment Garment, a pair of absorbent undershorts worn underneath the Liquid Cooling Garment.

Allegedly, none of the Apollo astronauts ever had to use their Fecal Containment Garments. The NASA report on Apollo waste management is careful to say things like "*should it have* become impossible for a crewman to have prevented defecation" and "*if* an uncontrolled bowel movement *had occurred*" (emphasis mine). In such an obviously drastic situation, the report concluded confidently, "the underpants would have contained the feces."

CHAPTER THREE

SPACE TOILETS

A decade after the first humans ventured into space for a few minutes at a time, both the US and the Soviet Union were ready to find out if people could live in orbit for weeks or months. The Soviet Union got there first with its Salyut 1 space station, which launched in 1971 and was equipped with the Solar System's very first space toilet.

On Earth, we rely on gravity to make sure our urine and feces go from our bodies into the toilet, and from the toilet bowl into the sewer system (or septic tank, depending on where you live). That's not an option in space. Instead, space toilets since Salyut 1 have used airflow from fans to create suction, which draws fluids away from astronauts and into waste management tanks. Astronaut Sally Ride once compared the resulting experience to "sitting on a vacuum cleaner."

Over the years, engineers have tried to make the toilets more efficient and comfortable to use, but the basic design hasn't changed.

HOW TO USE A SPACE TOILET

First, float your way into the "waste management compartment." Salyut 1 and Skylab were the first spacecraft big enough to have anything comparable to rooms, so the days of unzipping one's Constant Wear Garment to pee in the same seat where one

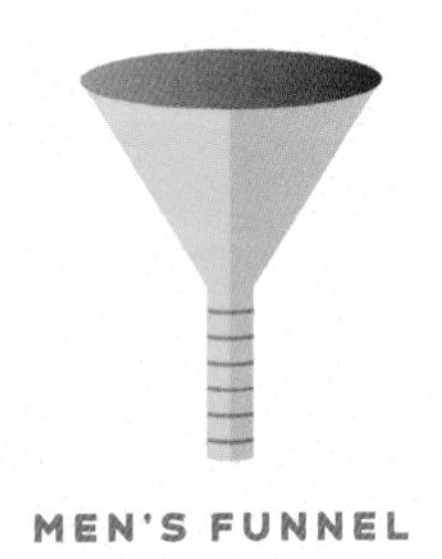

MEN'S FUNNEL

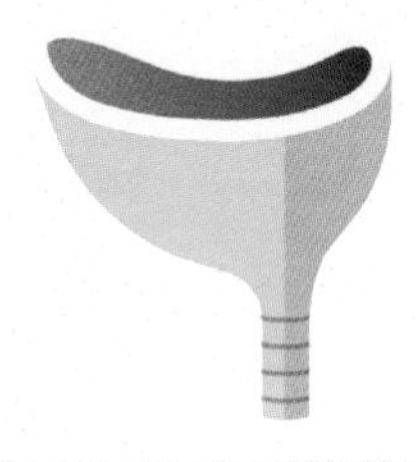

WOMEN'S FUNNEL

ate and slept were also over. For the first time, astronauts could go into a designated compartment and close a curtain or a flimsy panel. Of course, it was only about as private as a public restroom stall (maybe a little less), but it was better than Michael Collins having to roll on his Urine Collection Device in front of John Young.

If you're just there for a quick "number one," you select your personal funnel from the rack and connect it to the vacuum hose. Men's funnels look like, well, ordinary funnels. During training, NASA warns male astronauts not to put their personal equipment too far into the funnel, on account of the vacuum. In fact, there's no need to "hard dock," as astronauts put it, with the funnel at all.

Women's funnels are more oval-shaped, and they do require hard-docking; the funnel fits against the body, and it's important to get a good seal. Astronaut Rhea Seddon recommended having plenty of tissues on hand in case of funnel mishap.

"Over the course of a few days, you got to be reasonably good at it," she told space historian Amy Foster in 2011, "but it took some practice."

With the funnel in place, it's time to turn on the fan, open a valve, and fire away.

Pooping takes more finesse. You plant yourself firmly on the seat, using footholds and—on older models—thigh bars to hold yourself in place; floating away from the seat at the wrong moment would be a minor disaster, and the solution looks like safety restraints for the world's weirdest rollercoaster. A good seal is important, but so is positioning; you're aiming for a tiny four-inch hole, and if you miss, the mess is going to be pretty horrendous.

Once you're in position, you pull a lever to open the sliding cover over the hole, turn on the fan, and do your best.

"Not all astronauts are successful on the first try," wrote Jerry Linenger, who chalked it up to a combination of the weird setup and the body's adjustment to microgravity. "Many come out of the middeck toilet area mumbling 'another misfire.'" Of course, a misfire isn't the worst that can happen; if you forget to switch on the fan or break the all-important butt-seal with the seat, everything can just float back out of the commode in zero-G.

NASA'S SPACE TOILET SIMULATOR

Precise aiming isn't really an issue on Earth; we've got eighteen inches or so of toilet bowl, so if we're sitting on the seat, it's probably going in the right place. The toilets aboard the former Space Shuttle and aboard the ISS offer much smaller targets. Michael Mullane recounted having to practice "docking" with the four-inch-wide opening of the toilet's solid-waste transport tube during his astronaut training in 1979.

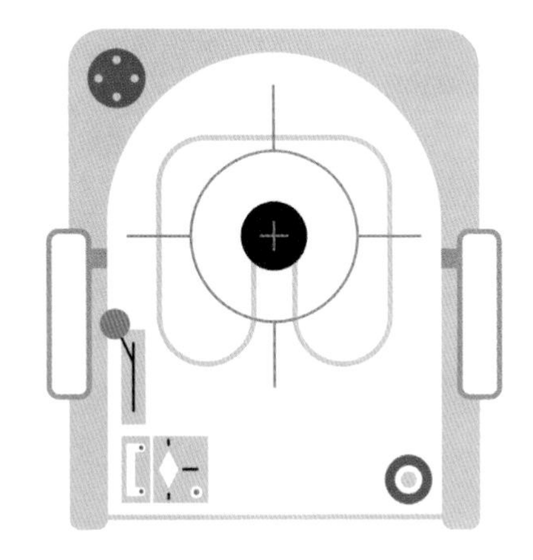

In a building at Johnson Space Center, next to the simulator astronauts used to learn to pilot the Space Shuttle in orbit, Mullane and his fellow Shuttle-era astronauts practiced perching on a mock-up space toilet. A camera inside the solid-waste tube projected a view of the astronaut's assets onto a monitor, complete with a crosshair to mark the exact center of the tube.

Astronauts fidgeted until they "were looking at a perfect bull's-eye," as Mullane put it. Then they made careful note of exactly where their thighs and seat were in relation to the toilet seat, hold-down clamps, and any other frame of reference they could find. If they missed, after all, the consequences could be dire.

ADVENTURES IN SPACE PLUMBING

If our toilet breaks, most of us run to the local hardware store for replacement parts or, in a worst-case scenario, call a plumber. In space, if something breaks, astronauts have to be resourceful enough to repair it themselves or improvise some other solution—or go without a toilet.

"The toilet is one of the pieces of equipment that gets a great deal of our attention," wrote Scott Kelly in his 2017 autobiography. That's not just because the alternative is uncomfortable; in an enclosed space like the ISS, where the air is cleaned and recirculated but never actually replaced by fresh air, keeping things sanitary is vital. If both toilets aboard the ISS broke down and couldn't be fixed for more than a few days, the crew would have to abandon the station.

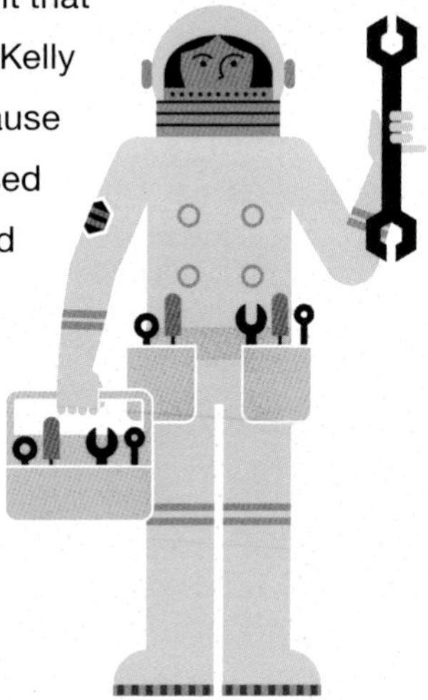

Additionally, astronauts depend on urine recycling for much of the water they drink aboard the ISS (more on that in Chapter Five). On future crewed missions to Mars, recycled urine may be the explorers' *only* source of water. The toilet will be critical life-support equipment.

"If we were on our way to Mars and the toilet broke and we couldn't fix it, we would be dead," wrote Kelly.

WHEN IT ALL GOES WRONG

On astronaut Judy Resnik's first flight aboard the Space Shuttle *Discovery* (1984's STS-41-D, not to be confused with the 1990 mission STS-41), the Shuttle's toilet failed halfway through the seven-day mission, forcing the crew to resort to backup options.

The male astronauts used leftover Apollo-era fecal collection bags and space sickness bags—with dirty socks stuffed inside to prevent splashing, after some slightly messy trial and error. The only woman astronaut on the flight, Resnik had trouble peeing into

the bags, so she cleverly improvised by holding a towel against herself, doing her business, then sealing the soaked towel into a bag for disposal.

If the mission had been scheduled to go on much longer, the crew would probably have had to cut it short. But with three days left to go, the resourceful astronauts managed to finish their week in space.

NO POWER, NO POOPING

Thirteen years later, Jerry Linenger and a crew of Russian cosmonauts found themselves in similar straits aboard the aging Russian space station Mir. The problem with a suction-driven toilet system is that it needs electricity to run. During Linenger's 1997 stint aboard Mir, the space station lost power several times, forcing astronauts to scramble to get it running again and then recharge batteries. While the batteries were recharging, Linenger and the cosmonauts had no electricity to spare for the fan, and they were forced to pee into a Russian version of the old-school Urine Collection Devices, rubber condoms and all.

"We all tried to avoid eating solid food," he added.

SCOTT KELLY VERSUS THE ACID BLOB

On his Space Shuttle flights and during his time on the ISS, Scott Kelly devoted a lot of time to maintaining the toilets. Usually that involved routine tasks like "compacting the toilet" or changing out parts. But once, aboard the ISS, Kelly had to do battle with a gallon-sized ball of bright purple pee and sulfuric acid.

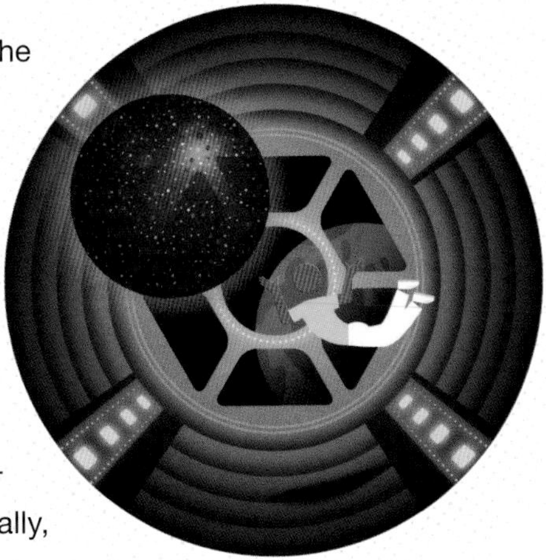

The toilet tank aboard the ISS mixes astronauts' urine with sulfuric acid to prevent the toilet from clogging and chromium oxide to prevent it from rusting. The chemicals also turn the astronauts' pee purple. And when a particular pump on the space station's toilet failed catastrophically, it unleashed "an enormous sphere of urine mixed with the sulfuric acid pre-treat" into the cabin.

Kelly described the floating purple sphere as the "creepiest" thing he ever dealt with in space.

TEST-DRIVING THE SPACE TOILET

Before sending the toilets into space aboard Skylab, NASA's engineers needed to make sure they'd designed a toilet astronauts could actually use in microgravity.

The only way to experience microgravity other than in space is on a plane in a steep dive, so NASA used a KC-135 airplane nicknamed the "Vomit Comet" (now retired) to train astronauts to maneuver in zero-G, one thirty-second dive at a time. And that meant the astronauts who test-drove Skylab's space toilet had to do their business—with an audience—in thirty seconds or less while the plane swooped toward the ground.

By 1972, NASA knew it planned to have women in its astronaut corps sometime within the next few years, so engineers had to make sure the toilet was practical for all astronauts. But NASA didn't have any women astronauts *yet*, so a group of Air Force flight nurses volunteered to help. At first, NASA engineers didn't even understand how women's urine was going to flow in microgravity given the difference in dispenser arrangements. The first group of flight nurse volunteers found themselves peeing during Vomit Comet parabolas while engineering crews filmed the urine streams.

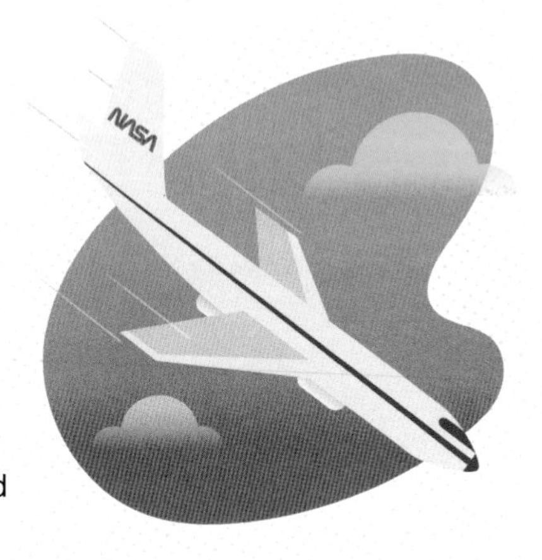

"Understandably, NASA's public affairs office did not announce this experiment publicly," wrote historian Amy Foster.

And a few years later, NASA's first group of women astronauts boarded the Vomit Comet to test the new urine funnels designed for women.

"You'd fill up your bladder before you left and hope there wasn't any delay," recalled Rhea Seddon. "Then when you're at zero-G, you pee. And of course, again, it's hard to pee on cue. And then if it starts to leak, if you don't have a good seal, you've got a full bladder and you're trying to stop."

THE NEW MODELS

NASA's latest and greatest space toilet design—which Artemis crews will use on the way to the Moon—is a $23 million project called the Universal Waste Management System (UWMS). The first UWMS was installed aboard the ISS in September 2020, and a slightly modified version will launch aboard the first crewed Artemis flight in 2024.

Engineers changed the shape of the funnel women astronauts use, and they also changed the positioning of both the funnel and the seat. The goal is to enable what astronauts call "dual ops" or "simultaneous evacuation," which means exactly what it sounds like: routing the urine and the feces into different tanks—for medical sampling and so the urine can be reprocessed into drinkable water (again, see Chapter Five). To prevent the results of "dual ops" from erupting out of the toilet in case an astronaut forgets to turn on the fan, lifting the lid or removing a funnel from its cradle automatically starts the fan, making for one less thing to remember.

The UWMS still uses sulfuric acid and chromium oxide to treat the urine, so for the foreseeable future, astronaut pee will always be purple.

CHAPTER FOUR

UNDER WHERE?

For astronauts stuck in their pressure suits—waiting for launch, spacewalking outside ISS, or exploring the south pole of the Moon—today's space toilets are as out of reach as Mercury-Redstone 3's liquid waste container was to Alan Shepard in 1961. And just like Shepard, astronauts still need a way to go when they're suited up.

NASA eventually opted for a very simple solution: astronaut diapers.

MAG
MAG

"DON'T YOU HAVE WIVES?"

The old Urine Collection Device is as much a thing of the past as the single-gender astronaut corps it was designed for; its condom-like attachment was (for obvious reasons) completely useless to women astronauts, who joined NASA in 1978. Of course, the women were also going to need a way to relieve themselves while in pressure suits, and NASA eventually settled on "just buy some adult diapers," but only after its engineers tried a more complicated approach first.

Their first attempt was a pair of briefs designed to hold a small funnel in place. In theory, the astronaut could simply do her business into the funnel, which—like the original Urine Collection Device—was connected to a plastic bag worn on a belt. The result was vaguely similar to the "sanitary belts" that once held women's menstrual pads in place, except with a funnel and a "vaginal seal" (the less said, the better) instead of an absorbent pad.

"It is absolutely as ridiculous as you can imagine," said National Air and Space Museum Space History Department Curator Jennifer Levasseur.

When Judy Resnik asked the NASA engineers involved—all men—whether they'd consulted their wives, the engineers seemed genuinely shocked by the idea. Although the first six women astronauts pretty much vetoed the whole design, NASA later tried to patent it as a healthcare device.

KEEP IT SIMPLE, SPACEFARERS

NASA eventually opted for a simpler solution. When Sally Ride became the first American woman to go to space in June 1983, she wore a pair of snug-fitting shorts with multiple absorbent layers and a plastic outer liner. Designed especially for spaceflight by NASA engineers, the undergarment featured a dry absorbent material that could soak up fluid and form a gel, which at least one early woman astronaut compared to sitting in a pile of JELL-O.

Disposable adult diapers were readily available in most drugstores by the early 1980s, much to the relief of people with various types of incontinence, but NASA engineers designed theirs specifically for spaceflight. The high-tech solution bore the suitably technical name Disposable Absorption Containment Trunks, or DACTs, which is definitely one way to spell "we don't want to call this a diaper." One of Sally Ride's (unused) DACTs is on display in the National Air and Space Museum.

A DIAPER BY ANY OTHER NAME

Ride's crewmates, meanwhile, were still stuck with the condom-and-bag setup that their forebears had worn to the Moon and back. On the one hand, the Urine Collection Devices had worked for more than two decades already, so there didn't seem to be much reason to change. And on the other, NASA engineers weren't sure how the men would respond to being told to wear diapers under their spacesuits.

"To be sure, no *Right Stuff*, red-blooded American astronaut would admit to wearing a diaper, so the personal equipment specialists came up with a different name: MAG," wrote Jerry Linenger in his 2000 autobiography. He went on to admit that he couldn't remember what MAG actually stood for, besides the G being for "Garment." (The answer is Maximum Absorbency Garment.) NASA started issuing MAGs to all its astronauts in 1988.

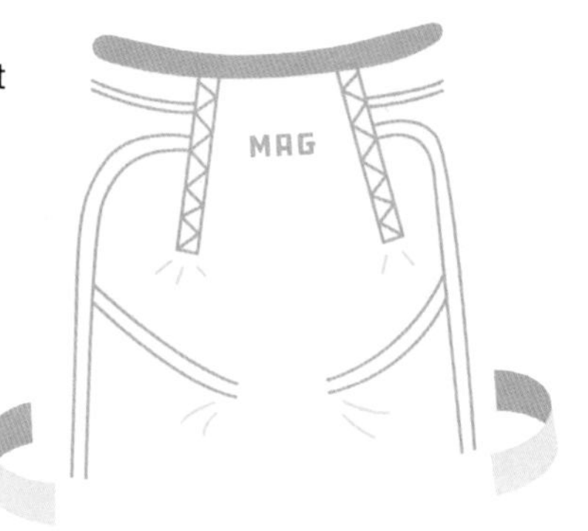

"The very early [MAGs] were designed very specifically by engineers at NASA or contractors," said Levasseur. "And eventually NASA finds out, as it has with virtually everything it's done for astronauts' interior cabin use, you can just use off-the-shelf stuff." Today's MAGs are commercial, off-the-shelf adult diapers designed for people with more down-to-Earth incontinence. They can hold about two liters of fluid for around eight to ten hours; that's near the top of the range for how much an average, well-hydrated person pees in a full day.

MAGs remain a staple of the spaceflight experience, but like everything else about relieving oneself in (or on the way to) space, using one takes some getting used to.

"It's hard to pee when you're lying on your back," astronaut Michael Foreman told *Air & Space Magazine* in 2011. "It's something psychological."

But apparently that's not the case for everyone. In his autobiography, Linenger described it as "almost impossible" to avoid the need to go when lying with one's feet above one's head for several hours. Most members of the Shuttle crews took advantage of their MAGs in the hours of waiting through the launch countdown.

"Whenever one of the crew began to whistle softly, we all knew what he was up to," shared Linenger.

THE SPACE POOP CHALLENGE

MAGs are a short-term solution, good for extravehicular activities (or EVAs, when astronauts venture outside the spacecraft) and for the few hours around launch or reentry. Beyond about ten hours, however, they're a soggy mess (and a rash or urinary tract infection waiting to happen), and they were never really meant to handle a solid number two. NASA launched the Space Poop Challenge in 2016, inviting members of the public to design something astronauts could use if an emergency left them stuck in their suits for days at a time. The goal? An under-the-suit setup that could handle everything an astronaut's body could dump into it, for up to 144 hours (that's six full days).

The winning design, courtesy of US Air Force flight surgeon Colonel Thatcher Cardon, proposed a tiny airlock in the crotch area of the spacesuit. Astronauts could use the airlock to change out their MAGs, or even inflatable bedpans. Most of the MAGs in use today are pull-on shorts, but Cardon's crotch-airlock design would require a switch to MAGs that fasten over the hips instead, to allow astronauts to change while still wearing a spacesuit.

"I mean, they can even replace heart valves now through catheters in an artery," Cardon told NPR in 2017. "So [the airlock] should be able to handle a little bit of poop!"

The second-place design uses airflow to move urine away from the astronaut and into a waste-storage compartment elsewhere in the suit. A third-place design also focused on moving the urine into holding sacks on the astronaut's thighs, where it gets disinfected and stored.

Shortly after the contest, NASA announced that it would eventually include elements of all three winning Space Poop designs in the gear that Artemis crews will wear to the Moon. But Artemis crews will also be wearing the tried-and-true MAG.

"We're still using diapers in the spacesuits," said Russell Ralston, deputy program manager for Extravehicular Activity at Axiom Space, during a March 2023 press conference after NASA unveiled its new Artemis Lunar Surface Suits (designed by Axiom). "They're just honestly a very effective solution. Sometimes simplicity is best."

CHAPTER FIVE

IT HAS TO GO SOMEWHERE

Now we know how astronauts actually get the job done, but what happens after? Once an astronaut has peed in a MAG or defecated into a bag, all that waste must go somewhere. In space, the answer hasn't changed much in sixty years: astronauts deal with bodily waste by either dumping it into space or hauling it back to Earth.

Oh, and sometimes they drink it.

CONSTELLATION URION

Gemini astronauts could empty their urine containment bags by connecting a valve on their pressure suits to the capsule's waste management panel and opening the aptly named "dump valve." Apollo astronauts could do the same thing after being suited up; the rest of the time, they connected their personal funnels directly to the panel and opened the valve. Either way, the urine got vented directly into the frigid vacuum of space through a tiny, filtered nozzle on the outside of the capsule. The urine dump nozzle was such a vital piece of equipment that engineers gave it a pair of small heaters—a primary and a backup—to prevent an icy clog that would shut down the whole system.

Once it cleared the nozzle, the astronaut urine instantly froze, forming a cloud of sparkling ice crystals that astronaut Wally Schirra nicknamed the "Constellation Urion." Michael Collins waxed poetic about the sight in his autobiography: "Cascading out in an irregular stream, they whiz past the window and tumble off into infinity, glistening virginal white in the sunlight instead of the nasty yellow we know them to be," he wrote.

Decades later, the Space Shuttles vented urine overboard in pretty much the same way. An especially large urine dump from Space Shuttle *Discovery* caused a minor stir among skywatchers in September 2009; the crew dumped about 150 pounds of urine and other wastewater, and the resulting trail of ice particles was visible from the ground.

In fact, it could be very difficult to tell Constellation Urion's glittering motes of astronaut pee apart from real stars, and both the capsules' guidance and navigation systems and the astronauts' eyes kept getting confused. Astronauts had to make sure they didn't schedule a urine dump and a navigational sighting within the same few-minute span, lest they end up sighting a sextant on a drifting droplet of crystallized urine.

ASTRONAUTS VERSUS THE PEE-CICLE

In September 1984, astronauts aboard the Space Shuttle *Discovery* had to use the Shuttle's robotic arm to dislodge a two-foot-long icicle of astronaut pee and other wastewater hanging from the ship's dump port. The urine was supposed to float away in a sparkling cloud, like always, but a stray particle of debris clogged the nozzle, and the urine clumped into a yellowish icicle. It stretched aft along the Shuttle's hull, past the edge of the open payload bay doors.

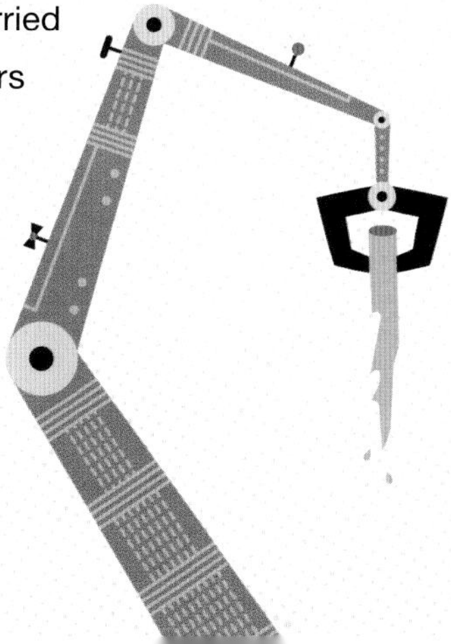

Astronauts and mission control worried that the pee-cicle could damage the doors if the crew tried to shut them, and that kind of damage could make it unsafe to return to Earth. Pilot Michael Coats rotated the Shuttle so the yellow icicle was in full sunlight for several hours. After some melting, Commander Henry Hartsfield gingerly used the robotic arm to knock the ice loose.

That flight was plagued by plumbing woes; it was the same mission on which the Shuttle's toilet broke down halfway through the flight (see Chapter Three).

"SMELLY WASTE!"

Number two has always been both easier and harder to get rid of than its counterpart. Aboard the Gemini and Apollo capsules, the carefully kneaded and rolled bags of poop went into a storage compartment (on which one crew member helpfully scrawled "Smelly waste!" with a marker) and came back to Earth with the crew.

On Space Shuttle flights, the toilet routed waste and urine into two separate compartments. Although gases from the fecal compartment vented into space (so the Shuttle occasionally trailed a faintly odorous miasma behind it), the fecal storage compartment had to be emptied back on Earth.

Today, aboard the ISS, fecal matter gets handled the same way it's been done since Mir: wet waste, as it's called, gets packed aboard a cargo ship (such as *Progress* or *Dragon*), along with accumulated garbage. When the uncrewed cargo ship returns to Earth, the heat and friction of passing through the planet's upper atmosphere turns the whole thing into a meteor of poop and garbage.

THE "TOSS ZONE"

The urine collection assembly in which Buzz Aldrin made history now lies on the regolith at Tranquility Base along with a pile of other things Aldrin and Armstrong literally tossed out of the Lunar Module at the last minute to save weight for liftoff.

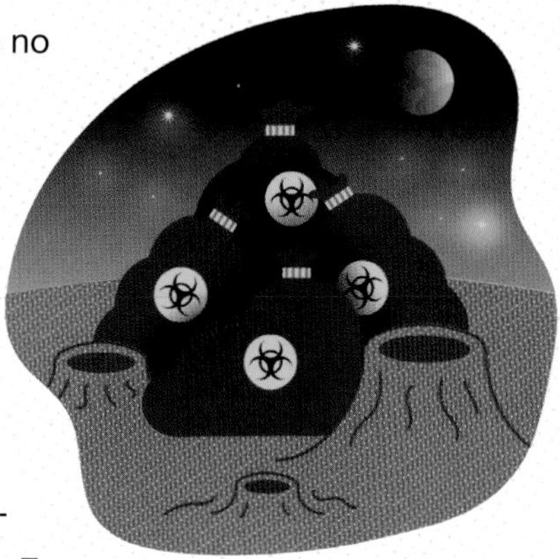

Lifting off the lunar surface was no sure thing, and the less weight the Lunar Module's engines had to lift, the better the astronauts' chances of making it back into space to rendezvous with the Command Module and fly home. And the astronauts had loaded up about fifty pounds of rock and regolith samples to bring back to Earth. For about eight minutes before lifting off, Aldrin and Armstrong stood on the ladder of the Lunar Module and tossed out anything they wouldn't need for the return trip—various scientific instruments and tools, reference books, the plastic bag the American flag had been wrapped in, and some used fecal collection bags and urine collection assemblies.

The Lunar Legacy Project at New Mexico State University has a very thorough list of everything the crew left behind, along with a map of what's now called the Toss Zone. Archaeologist Beth O'Leary and her colleagues say the pile of objects Aldrin and Armstrong left behind are part of humanity's first archaeological site in space. Archaeologists on Earth can learn a lot about a civilization by rummaging through its garbage heaps, and someday archaeologists may piece together details of life during the Space Race by examining Aldrin's used Urine Collection Device.

"TODAY'S COFFEE IS TOMORROW'S COFFEE"

Astronaut urine meets a very different fate today than it did a few decades ago. As astronaut Jessica Meir cheerfully quipped to the media, "Today's coffee is tomorrow's coffee!" Drinking—and even breathing—recycled urine has been a feature of life in space since the final days of the Mir space station, and today it's crucial to survival aboard the ISS.

Jerry Linenger reports that during his 1997 stint on Mir, he and his cosmonaut shipmates agreed that they could handle the idea of drinking their own pee to survive, but not each other's. But they had no problem breathing oxygen reclaimed from each other's urine. So, they chose to put the reclaimed water through more processing, which split the water molecules into hydrogen and oxygen (sort of the reverse of the process Apollo 11's fuel cells used to *make* water from hydrogen and oxygen).

Astronauts aboard the ISS today don't have the luxury of squeamishness about drinking their crewmates' urine. About 90 percent of the station's water comes from a urine recycling system installed in 2008. During his eleven-month stint on the station in 2015 and 2016, Scott Kelly drank about 193 gallons of recycled pee. A lot of that pee, Kelly later said, came from the Russian side of the station: "Cosmonaut urine is one of the commodities in an ongoing barter system of goods and services between the Russians and the Americans," he wrote.

NASA hopes that the first people to fly to Mars will get 98 percent of their water from urine recycling. Water is heavy—the second-heaviest thing a rocket carries after fuel. Launching enough water into space to sustain astronauts all the way to Mars and back (not to mention building room into a spacecraft to carry it all) would be staggeringly impractical. Mars missions are going to have to be more self-sustaining than missions to low-Earth orbit or even the Moon.

Even if it means astronauts living on their own urine.

CHAPTER SIX

FARTING AROUND

The average human farts between five and fifteen times a day, whether they're on Earth or trapped in a small spacecraft cabin with a captive audience. Bacteria in the colon help finish off the process of digestion, and those bacteria exhale a mixture of carbon dioxide, hydrogen, methane, nitrogen, oxygen, and sulfur as a by-product. And like the other products of digestion, it's got to go somewhere.

Despite fire hazards and bad smells, astronauts find farting as hilarious as everybody else in the Solar System.

CAN YOU PROPEL YOURSELF BY FARTING IN SPACE?

Alas, even after a hearty meal of beans and cabbage, you couldn't propel yourself around the International Space Station by what Jerry Linenger called "digestive gas thrusters."

The human body just doesn't produce enough gas or expel it with enough force (no matter what sometimes seems to be happening down there). There are about thirty-six ounces of gas in the average person's typical fart, and it exits the body at about 6.8 miles per hour. That's not enough force to shift the mass of a human body much, even in microgravity. After all, the International Space Station is full of air, which creates friction and slows things down.

But, like true heroes of space exploration, several astronauts have tried to fart around and find out:

- "We all tried it," admitted Canadian astronaut Chris Hadfield in a Reddit question-and-answer session, but the human body comes equipped with "not the right propulsive nozzle."
- "It's not as propulsive as you would think," US astronaut Mike Massimino told *Gizmodo* in 2017. "It's easier said than done."
- "In what I thought was a real voluminous and rapidly expelled purge, I failed to move noticeably," US astronaut Roger Crouch told author Mary Roach in *Packing for Mars*.

However, it's apparently very possible to bounce yourself off a space station toilet enough to break the all-important butt-to-seat seal. And that's not the only flatulence-related hazard in space.

FIRE HAZARD

As a quick YouTube search will confirm, flatulence is flammable. That's mostly thanks to methane and hydrogen. And in an enclosed space with dubious ventilation and no escape, flammable gas is a very dangerous prospect.

Early in the American space program, NASA restricted astronauts' in-flight menus in an effort to keep them as gas-free as possible. Beans, cabbage, and other gas-inducing foods like sprouts were off-limits in those early years.

In 1964, a US Department of Agriculture scientist named Edwin Murphy offered NASA another solution—just recruit less gassy astronauts. After a series of experiments, he estimated that about half the US population didn't produce any methane in their digestive tracts. These methane-free people still passed gas, but the result was much less flammable. Murphy argued that NASA should recruit methane-free astronauts for the space program.

NASA decided it was easier to limit gassy foods than to recruit astronauts based on the contents of their flatulence. Even those requirements have relaxed over the decades, thanks in part to better ventilation systems and more room on spacecraft like the Space Shuttle and the ISS. Today, astronauts aboard the ISS can eat pretty much whatever they want, even cabbage grown aboard the station.

Russian cosmonauts never bothered with anti-flatulence diets. Cabbage and borscht have been staples of their space diets for as long as there have been cosmonauts.

BUBBLY WATER ON APOLLO 11

After their historic landing on the Moon, the crew of Apollo 11 flew home in triumph and a smelly cloud of flatulence.

The problem was the water. Apollo 11's fuel cells combined hydrogen and oxygen in a chemical reaction that generated enough energy to power the spacecraft. The reaction also produced water as a convenient bonus, but with some extra hydrogen bubbles in the mix. That sounds festive and fun, but it turned into some especially noxious farts.

Michael Collins later described the smell as "a not-so-subtle and pervasive aroma which reminds me of a mixture of wet dog and marsh gas."

THE FIRST FART ON THE MOON

The first person to fart on the Moon (or at least the first to admit it on a publicly broadcast channel) was astronaut John Young during the Apollo 16 mission in 1972.

"I have the farts again," he radioed to Lunar Module pilot Charles Duke. "I got them again, Charlie. I don't know what the hell gives them to me." However, Young did have his suspicions: the potassium-fortified fruit juice NASA had added to the astronauts' space diets. It's true that fructose (fruit sugar) gives some people gas, and Young may have been one of the unlucky ones. During his radio call about "the farts," Young vowed never to eat another orange.

SENDING "EMAILS" TO THE ISS

Even without fire to worry about, nobody wants to be trapped in a small, stuffy spacecraft cabin with the gassy remnants of everyone's latest meal or drink. That's why astronauts onboard the ISS try to quickly float to the restroom when they need to let one rip, according to Mike Massimino. The station's toilet area has excellent ventilation, for obvious reasons.

"Either you do it in private or get people mad at you," Massimino said. "That's the kind of thing that can lead to crew disharmony."

Of course, some shenanigans are too funny to resist, even for professional astronauts in space. Astronaut Clayton Anderson, in a post on Quora, confessed that he once positioned himself at the end of an air hose leading from the middeck of the Space Shuttle to the ISS docking tunnel and "fired away."

"We started referring to these events as 'sending emails to the ISS,'" he wrote.

WHAT HAPPENS IF YOU FART IN YOUR SPACESUIT?

Not much, it turns out. If a single fart, or even a spate of them, can't budge an astronaut in normal clothes, it certainly can't do much inside a spacesuit, so there's no danger of spacewalking astronauts tumbling out of control thanks to errant digestive gas.

Modern spacesuits have constant airflow inside, which dissipates any noxious gases before they reach the astronaut's nose. The suits also carry a filter made of lithium hydroxide, which is designed to clear away the carbon dioxide astronauts exhale, but it works just as well on methane.

And unless the astronaut decides to share the news—like John Young—what happens in the spacesuit, stays in the spacesuit.

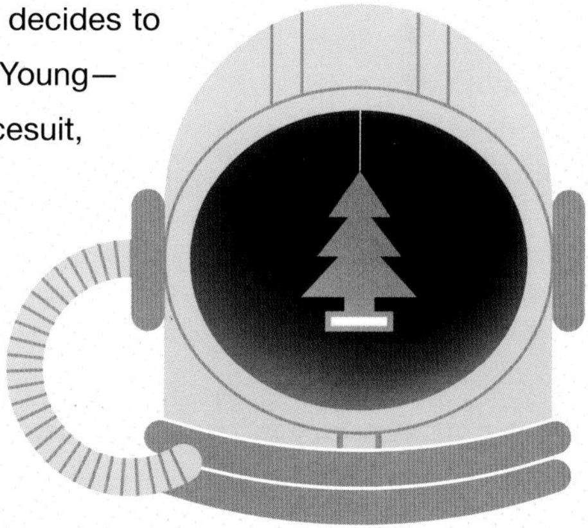

NASA
SICKNESS

CHAPTER SEVEN

OUT THE WRONG END

Not all the bodily waste astronauts deal with comes out the aft end of their spacesuits. The human digestive system, like any other piece of equipment, sometimes malfunctions, and when that happens, partially digested material tends to come out the same way it went in. Usually, that's not a big deal. Modern astronauts are prepared with handy emesis bags and anti-nausea pills, and they're also taught to expect space sickness as a routine part of flight.

THE FIRST PERSON TO VOMIT IN SPACE

In the early years of human spaceflight, space sickness was a worrisome mystery. It started with cosmonaut Gherman Titov, the second human ever to go to space. He blasted off aboard Vostok 2 in August 1961, a few months after Yuri Gagarin's historic flight. A few hours—and a few laps around the planet—later, Titov felt queasy and became the first person to vomit in space.

At the time, neither NASA nor the Soviet space program fully understood how the human body would react to being in space. Titov's illness alarmed Sergei Korolyov, the enigmatic engineer in charge of the Soviet space program (known, even to the cosmonauts, only as the "Chief Designer"), enough to put the country's whole crewed spaceflight program on hold for a year while doctors diagnosed the problem.

The answer turned out to be space sickness, or space adaptation syndrome, in more clinical terms. Humans evolved in a world where "up" and "down" weren't abstract concepts, and our bodies and brains constantly keep track of how we're oriented in relation to the ground. Some of that information comes from visual cues, and some of it comes from the motion of fluid in our inner ears. And when those clues conflict—like when an astronaut is floating weightlessly in the cabin of a spacecraft, looking up at what they thought was the floor—the result can be messy.

THE MYSTERIOUS RUSSIAN SICKNESS

For several years after poor Titov's historic heaves, NASA insisted that space sickness was just a Russian problem.

"We Americans, however, had tended to pooh-pooh these reports," wrote Collins in his autobiography. "We based our own asymptomatic flight experience upon the fact that we were flying only experienced test pilots, whose inner ears were accustomed to such buffets and bumps."

And then Frank Borman, the mission commander of Apollo 8, got sick.

In December 1968, the crew of Apollo 8 was on its way to the Moon when Borman was suddenly ill. The bout of nausea came seemingly out of nowhere, and Borman radioed Mission Control and asked to speak privately with medical staff. But the astronaut couldn't shed much light on what was wrong; feeling horrible and throwing up could be symptoms of anything from a deadly virus to the aftereffects of a sleeping pill Borman had taken the day before launch. Nobody wanted to turn around and cancel the mission, least of all Borman, so he tried to wait it out.

And within a day or two, Borman was back on his feet—figuratively, anyway.

Borman turned out to be the first American astronaut to suffer from a zero-gravity ailment the Soviet space program had been describing for years: space sickness.

Eventually, NASA realized that American astronauts hadn't gotten space sick until Apollo 8 not because they were tougher or more experienced than their Soviet counterparts, but because American spacecraft were smaller. Mercury and Gemini crews barely had room to get out of their seats, which meant they didn't have room to float around and experience the mixed-up orientation cues of weightlessness. The Apollo command module was much larger, and for the first time, astronauts had room to float around and feel the full, sometimes nauseating, effects of microgravity.

Soviet Vostok capsules, on the other hand, had been roomy enough for space sickness to set in. Gagarin probably escaped it only because his flight was so short. Ironically, if Titov had spent more than a day in orbit, his space sickness would probably have passed, leaving him free to enjoy the experience.

A COMMON PROBLEM

Today, scientists estimate that more than three-quarters of people are prone to space sickness. It doesn't matter if you're a seasoned test pilot, an avid scuba diver, or a rookie space tourist. It doesn't even seem to matter if you're prone to motion sickness; the only factor that seems to predict a person's odds of suffering from space sickness is whether they got space sick the last time they left the planet. First-timers in space just have to roll the dice.

The good news, according to Jerry Linenger, is that only about 10 to 20 percent of space sickness sufferers actually have to use the space sickness bag NASA tucks into the front leg pocket of astronauts' spacesuits. Like everything else NASA sends to space, the space sickness bag is a tiny feat of engineering. It even comes with a built-in cloth flap for wiping the sick astronaut's face and a Velcro fastener to stick to the nearest wall.

Fortunately, the human body just needs time to adjust to the sensation of moving around in microgravity.

"By the second day in Earth orbit, practically everyone feels better," wrote Linenger.

CHUNKY BUBBLES

You can burp in space, but just because you can doesn't mean you should.

And you probably don't want to. Trust me.

Here on Earth, burping is the body's way of getting rid of air in the upper digestive tract. The air floats, while all the solid chunks and the dribbles of liquid left over from your last meal sink to the bottom of your stomach. When you burp, the air comes out, and the rest stays down. (In theory; no system is perfect.)

In space, however, everything floats, so the partially digested remnants of your last meal are all mixed up with the air you've managed to swallow. Chris Hadfield described it, a little too evocatively, as "chunky bubbles." And that means that when you burp, everything comes out: air, liquid, and solid.

Supposedly, if you really need to burp, you can push yourself off a wall, according to astronaut Jim Newman and author Ariel Waldman. For a second or two, that acceleration will work like artificial gravity inside your body, sorting out the chunky bubbles in your stomach into their constituent parts. For a brief moment, you can burp safely, but you better be quick about it.

EPILOGUE

HANGING UP THE FUNNEL

At first glance, this topic might have seemed like a silly one for a book . . . and, on one hand, it is. That's been half the tremendous fun of writing it.

I'd be violating my journalistic integrity if I pretended not to enjoy the daylights out of a good fart joke. Say, did you hear the one about Uranus?

(It smells like methane. Really.)

But, on the other hand, "how do you go to the bathroom in space?" is a totally valid question, and I hope the answer has made you think about the history of space exploration a little differently. It's one thing to memorize the date of the Moon landing and the names of the Apollo 11 astronauts; it's another to realize they set off for the Moon in a cramped space capsule where their best option for relieving themselves involved pooping in a plastic bag. The whole story feels more real, and more like a wild adventure, when you know some of the weird, messy details.

Ultimately, that's what space exploration is: an adventure—maybe humanity's greatest ever.

A TIMELINE OF PEEING AND POOPING IN SPACE

APRIL 1961: First human in space (Yuri Gagarin)

MAY 1961: First American in space (Alan Shepard)

AUGUST 1961: First case of space sickness (Gherman Titov)

FEBRUARY 1962: First American orbits the Earth (John Glenn)

FEBRUARY 1962: First person known to pee in space (John Glenn)

JUNE 1963: First woman in space (Valentina Tereshkova)

DECEMBER 1968: First known American case of space sickness (Frank Borman)

DECEMBER 1968: First humans orbit the Moon (Frank Borman, James Lovell, and William Anders)

JULY 1969: First human steps on the Moon (Neil Armstrong)

JULY 1969: First person pees on the Moon (Buzz Aldrin)

APRIL 1972: First known person to fart on the Moon (John Young)

JUNE 1983: First American woman in space (Sally Ride)

JUNE 1983: First woman known to pee in space (Sally Ride)

SUGGESTED FURTHER READING

Aldrin, Buzz, with Wayne Warga. *Return to Earth*. New York: Random House, 1973.

Collins, Michael. *Carrying the Fire: An Astronaut's Journeys*. New York: Farrar, Straus and Giroux, 1974.

Foster, Amy. *Integrating Women into the Astronaut Corps: Politics and Logistics at NASA, 1972–2004*. Baltimore: Johns Hopkins University Press, 2011.

Kelly, Scott. *Endurance: A Year in Space, A Lifetime of Discovery.* New York: Knopf, 2017.

Linenger, Jerry. *Off the Planet: Surviving Five Perilous Months Aboard the Space Station Mir.* New York: McGraw Hill, 1999.

Mullane, Mike. *Riding Rockets: The Outrageous Tales of a Space Shuttle Astronaut*. New York: Scribner, 2006.

O'Leary, Beth, Lisa Westwood, and Milford Wayne Donaldson. *The Final Mission: Preserving NASA's Apollo Sites*. Gainesville: University Press of Florida, 2017.

ACKNOWLEDGMENTS

Peeing in space happens in a vacuum, but writing a book doesn't. You wouldn't be holding this book in your hands without a whole crew of amazing people.

First and foremost, this book wouldn't exist if my editor, Randall Lotowycz, hadn't come up with the idea. I'm extremely grateful to him for offering me the chance to write it, and for shepherding it through all the steps that came after.

A huge thank you to Paul Kepple at Headcase Design for his amazing illustrations and Katie Benezra for her brilliant design work, as well as production editor Amber Morris, copyeditor Duncan McHenry, proofreader Zan Ceeley, associate editor Britny Perilli, and everyone else at Running Press for seeing this through its production.

Thanks are also due to Jennifer Levasseur at the National Air and Space Museum for answering my frantic last-minute questions about exactly how those roll-on cuffs worked, and to Beth O'Leary at New Mexico State University for sparking my interest in the Apollo 11 Toss Zone.

To my incredible editors at *Inverse*, John Wenz and Claire Maldarelli, and my editor as a freelancer at *Ars Technica*, John Timmer, thank you all for your guidance and your patience. I wouldn't be at the book-writing stage of my career without you all. And to all my *Inverse* colleagues, thank you for being completely awesome to work with and for constantly inspiring me to be a better journalist and a better writer. Y'all are all brilliant, and I'm lucky to work with you.

Daddy, I wish you could have been here to see this happen, but your memory was with me every step of the way. Mom, thanks for literally everything, but especially for years of encouragement and for loving me even on the days I barely even like myself (and for all the sparkly pens).

KIONA N. SMITH has spent the last decade telling cool stories about science on the internet, and now they've written a book. Kiona is a space reporter at *Inverse* and a freelance archaeology reporter at *Ars Technica*. Based in Tulsa, Kiona shares an office with a scruffy little dog, a very jumpy gecko, and an ever-growing pile of unfinished knitting projects. Find them online at linktr.ee/kionasmith.